Amelia Earhart

by Jill C. Wheeler

A&D BIOGRAPHIES
Breaking Barriers

visit us at
www.abdopub.com

Published by ABDO & Daughters, an imprint of ABDO Publishing Company, 4940 Viking Drive, Suite 622, Edina, Minnesota 55435.

Printed in the United States.

Edited by Paul Joseph
Graphic Design: John Hamilton
Cover Design: Mighty Media
Photos: Corbis and AP/Wide World

Library of Congress Cataloging-in-Publication Data

Wheeler, Jill C., 1964-
Amelia Earhart / Jill C. Wheeler
p. cm. -- (Breaking Barriers)
Includes index.
Summary: Traces the life of the pilot who became the first woman to fly across the Atlantic Ocean and mysteriously disappeared in 1937 while attempting to fly around the world.
ISBN 1-57765-318-1
1. Earhart, Amelia, 1897-1937--Juvenile literature. 2. Women air pilots--United States--Biography--Juvenile literature. [1. Earhart, Amelia, 1897-1937. 2. Air pilots. 3. Women--Biography.] I. Title. II. Series.

TL540.E3W44 2002
629.13'092--dc21
[B]

98-021868

Contents

Women Can Have Careers, Too

The Purdue University dining room hummed with its usual conversation one evening in the fall of 1935. This evening there was a special guest. She was a tall, slender woman with short brown hair and a smiling face. As the other diners straggled back to their dormitory rooms, many young women stayed to speak with this guest.

"If you want a certain job, try it," the woman told the students. "If you should find that you are the first woman to fuel an urge in that direction, what does it matter? Fuel it and act on it just the same."

EARHART
NAUTIQUE
ONAUTIC
U.S.A.
No. 6017
ciation, recognized
nautique Interna-
authority for the
certifies that
nditions required by
tique Internationale,
s hereby brevetted
OMMITTEE
Chairman

This was a new concept for many of the students. In those days, most people expected women to get married and stay at home raising a family. Yet this visitor was encouraging them to become doctors, scientists, engineers, or whatever they dreamed of being.

Some people on the Purdue campus thought this woman was a threat. What if the female students listened to her? What if they chose not to get married and have careers instead? The woman only laughed at their fears. Women could have careers, or a marriage and careers, she said. She had managed to combine them both.

When not counseling, the woman often could be found at the Purdue airfield. Her passion was flying airplanes. She had become the first woman to fly across the Atlantic Ocean, both as a passenger and as a pilot. She had set many other aviation records as well.

For this visitor was no ordinary guest lecturer, she was Amelia Earhart.

Amelia Earhart is shown in pilot's headgear after applying for a pilot's license.

A Privileged Beginning

Amelia Mary Earhart was born on July 24, 1897, in Atchison, Kansas. She was the daughter of Edwin and Amy Earhart, yet she was born in the home of her mother's parents, Alfred and Amelia Otis. The Otis family wanted Amelia's mother to have her child there because she had suffered a miscarriage earlier.

The Otis family was wealthy and influential. Alfred Otis always believed his daughter Amy, (Amelia's mother) had married beneath her social status when she chose Edwin. In fact, Amy's father forbade her to marry him until he was earning at least $50 a month.

Edwin was the son of a minister. He earned a law degree by working odd jobs while attending the University of Kansas. After graduation, he took a job with the railroad. The job meant he often was away from home. Amy frequently traveled with him. That left Amelia and her little sister, Muriel, to stay with their grandparents.

Amelia Earhart and her mother, Amy.

Amelia Earhart as a young girl.

Amelia and Muriel had an unusual childhood. They did not always behave the way little girls were supposed to behave in those days. They wore funny looking pants, called bloomers, when most girls wore dresses. They played baseball and went fishing when other girls were learning how to sew. They loved to play outside when other girls spent most of their time inside. "I am sure I was a horrid little girl," Amelia said of herself later.

Amelia loved to go fast. When she was a little girl, she and some friends built a roller coaster. It was a creaky track running to the ground from the top of a woodshed. The car was a box with planks on the bottom. Amelia was the test pilot. She crashed the box into the ground several times before she got it right. She said it was just like flying.

Amelia's mother, Amy, had some of her daughter's spirit. She had been the first woman ever to climb to the top of Pike's Peak in Colorado. The guides who accompanied her were amazed any woman could do that.

Amelia saw her first airplane when she was 10. She and her family were at the Iowa State Fair. The plane was a wooden biplane that looked much like a large box kite. The strange-looking plane didn't hold Amelia's interest. "It was a thing of rusty wire and wood and not at all interesting," she recalled later.

On the Move

In 1909, the Earhart family moved to Des Moines, Iowa. Amelia's father had taken a job as head of the Rock Island Railroad's claims department. The job meant he and his family had their own private railway car. The Earharts traveled in the car to Kansas City, Chicago, and all the way out to California. Amelia loved to travel.

The Earharts lived in Des Moines for three years. During that time, Edwin developed a problem with alcohol. His drinking problem finally cost him his job. No one wanted an employee they could not depend upon because he might be drunk. He had to take a different job as a clerk for the Great Northern Railway. The change forced the Earhart family to move to St. Paul, Minnesota.

That position was short-lived as well. Once again, Edwin's drinking cost him a job. The next job he found was in Springfield, Missouri. Amy, Amelia, and Muriel moved with Edwin to Missouri. When they arrived, they learned there was no job for Edwin after all. Frustrated, Amy moved to Chicago with her daughters to stay with friends. Edwin returned to Kansas City. He lived with his sister there while he worked to build a law practice.

Amelia graduated from Hyde Park High School in Chicago in 1916. Hyde Park was the sixth high school she had attended. Despite many moves, Amelia was determined to finish high school on time. The students there knew her as a loner. That was because one of her teachers had a hearing problem. Amelia tried to defend the teacher. After that, many of the other students didn't want anything to do with her.

Amelia Earhart in 1916 after graduating high school.

Following Amelia's graduation, the family reunited in Kansas City. Edwin had managed to overcome his alcohol problem and was working successfully as a lawyer.

Amelia's next adventure was to enroll at the fashionable Ogontz School in Pennsylvania. She could attend the expensive school because her mother had inherited some money. While there, she kept a scrapbook of clippings. The clippings told the stories of women who had carved out careers in fields dominated by men. There were women working as brick layers, bank presidents, fire lookouts, and more. Amelia liked the idea of succeeding in a non-traditional field.

Amelia only attended Ogontz for a little while. In December 1917, she visited Muriel, who was in college in Toronto, Ontario, in Canada. The sisters were out walking one day when they saw several one-legged men on crutches. The men were soldiers wounded in World War I. "There for the first time I realized what the World War meant," Amelia said later. "Instead of new uniforms and brass bands, I saw only the results of four years' desperate struggle; men without arms and legs, men who were paralyzed and men who were blind... ."

Suddenly, the war meant much more to Amelia. She decided to quit school and help with the war

A daredevil performs a wingwalking stunt.

effort. She became a nurse's aid at the Spadina Military Hospital in Toronto.

Many of the patients Amelia nursed were military pilots. They told her stories of their adventures and air combats, called dogfights, during the war. Amelia listened intently to their stories.

One day when she wasn't working, Amelia ventured to the Toronto Fair to watch stunt pilots perform in their brightly colored biplanes. They amazed the crowd with their daredevil antics. Amelia became curious about how those airplanes worked. She signed up to take a class in automobile engine mechanics to learn.

Catching the Flying Bug

World War I ended in 1919. Amelia ended her job as a nurse then as well. She moved to New York City to study medicine at Columbia University. She was a good student, yet something in her life was missing. She never could forget the pilots she had seen and met. She spent a year at Columbia, then moved to Los Angeles, California, to be with her parents. They were having problems with their marriage and Amelia wanted to help.

Amelia had planned to finish her medical studies in California. That changed after she and her father visited an air show there. She took a 10-minute ride in an open-cockpit biplane over Los Angeles. "As soon as we left the ground, I knew I had to fly," she said.

She signed up for flying lessons. Yet learning to fly was expensive. Amelia took a job as a file clerk for a telephone company to pay for her lessons. Whenever she had time, she took the streetcar to the end of the line. Then, she walked several more miles to a dusty airfield. She also began to dress like a pilot. She cut her hair short and wore khaki pants, a leather jacket, boots, and goggles.

When Amelia took flying lessons, she started dressing like a pilot.

Amelia received her pilot's license in 1922.

Amelia earned her pilot's license in the summer of 1922. That made her one of only about a dozen licensed women pilots in the world. She could stop paying for lessons, but she still had to rent the airplane. She took on a variety of odd jobs to pay for plane rentals. Once she even drove a truck for a gravel company.

Her family came to her rescue and bought her a plane for her 24th birthday. It was a yellow biplane called a Kinner Canary. Now she could fly whenever she wanted. She set out to work perfecting important skills. She learned how to fly her plane out of stalls and spins. She even learned how to land her plane when the engine quit.

All of these were critical skills. Airplanes in the 1920s were far less reliable than they are today. Pilots often spent more time repairing their planes than they spent flying them. Amelia also learned how to fix her plane.

Amelia began her own stunt flying in the fall of 1922. She gave her family tickets to an air show without telling them she would be in the show. Her stunt set a new women's altitude record. She flew her *Canary* up to 14,000 feet (4,267m) before the engine quit. Fortunately, she knew how to land without an engine. Her record only stood for three weeks, but she had taken the first step toward making a name for herself in aviation history.

No one paid Amelia to fly, so she still had to work at other jobs. Despite those jobs, she couldn't afford to keep her bright yellow airplane. She had to sell it and stop flying. Then her parents divorced. Amelia bought a yellow car instead and drove across the United States to move herself and her mother to Boston, Massachusetts.

The *Friendship*

In Boston, Amelia took a job as a social worker at a settlement house for immigrants. She still felt strongly about helping other people. Her job was to work with the mostly Chinese and Syrian immigrants and their children. She helped them learn English and get settled in their new country. She loved the job, but she desperately missed flying.

There was a lot happening in aviation at the time. In May 1927, the world practically held its breath as Charles Lindbergh became the first person to fly across the Atlantic Ocean alone. He flew from New York to Paris in just under 34 hours. Shortly after Lindbergh's flight, an English woman decided a woman should make that flight as well. She bought a plane, hoping to do it herself. However, her family talked her out of it. They said it was too dangerous. Three woman already had died trying to cross the Atlantic. So the woman set about finding a different woman to do the job.

One day Captain Hilton Railey called Amelia and asked if she would be that woman. Unlike Lindbergh, Amelia would not be doing the flying.

She would simply be a female passenger. Pilot Wilmer Stultz and a mechanic named Lou Gordon would be with her. A panel of promoters for the flight interviewed Amelia and decided she would be perfect. She even looked a little like Lindbergh.

Amelia, Stultz, and Gordon began to prepare for their flight. They would be flying in a bright orange Fokker Trimotor called the *Friendship*. The orange color would make it easier to find the plane if it went down in the water. The plane also had pontoons to help it float.

Despite the precautions, Amelia knew the journey would be dangerous. Radio and navigation equipment was primitive. There was no reliable weather information. Nor was there deicer to keep ice off the plane's wings in cold weather.

Amelia stands in the doorway of the Friendship.

The crew kept the flight a secret until after they had taken off. They didn't want anyone else trying to set the record first. The *Friendship* was delayed 12 days by weather. Finally, they took off on June 17, 1928, from Newfoundland, Canada.

Four hours into the flight they hit fog. That evening, their radio went dead. They had to navigate by compass. At dawn, they did not know where they were. At last, Amelia spotted an ocean liner. The *Friendship* crew tried to reach the ship by radio—no response. Then they circled it, hoping the ship would paint its position on the deck as ships often did for airplanes. Still no luck. Amelia tried to throw the ship a note attached to an orange, but she missed and it landed in the water.

With no luck getting help, the crew guessed the right direction and kept flying. They were down to one hour of fuel left when they sighted land. They set the plane down in the water near Burry Point, Wales. Amelia had become the first woman to fly across the Atlantic.

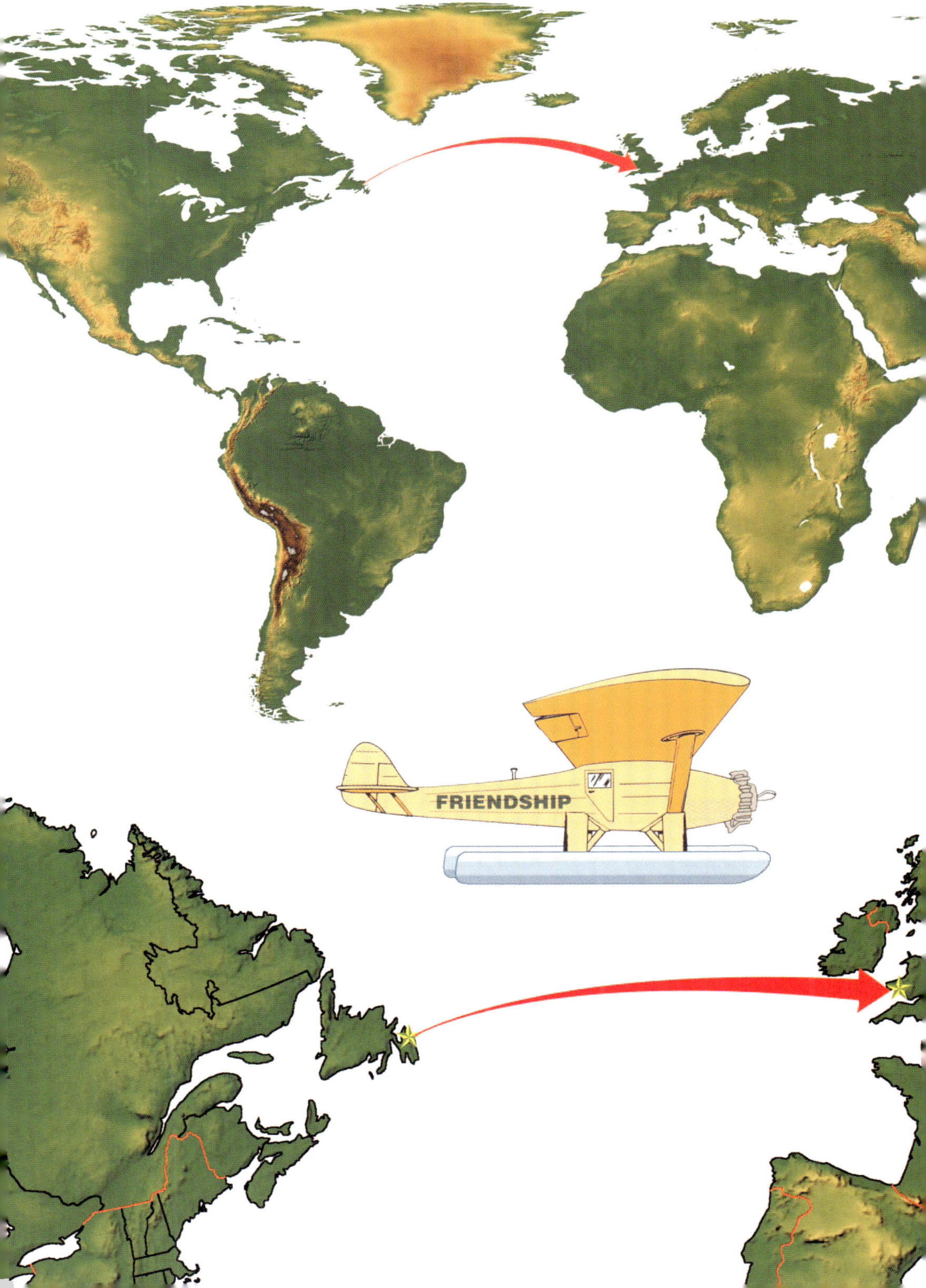
FRIENDSHIP

A New Celebrity

Suddenly, Amelia was a celebrity. Everyone wanted to see the first woman to fly across the Atlantic. The attention made Amelia uncomfortable. She hadn't been the pilot, just a passenger. She kept giving the credit to the real pilot. She decided that someday she would earn the honor by flying across the Atlantic all by herself.

By now Amelia had a manager named George Putnam. Putnam had been one of the promoters for the *Friendship* flight. His family was in the publishing business, so he asked Amelia to write a book about the flight. He even offered that she could stay with him and his wife while she wrote. Amelia accepted the offer and the peace and quiet it gave her.

After she finished her book, *20 Hrs., 40 Min.*, she again felt the need to fly. The book, titled after the time it took to make the journey, also gave her enough money to buy another airplane. This time, she bought a lightweight Avro Moth and headed west.

Amelia's book about her transatlantic flight was called 20 Hrs., 40 Min.

Following roads, railroad tracks, and maps, she flew to California, then turned around and flew back. She didn't know she had just become the first woman to fly from the Atlantic to the Pacific and back.

Amelia traveled around the country, speaking to people about women and flying.

Amelia's celebrity status continued upon her return. People around the country were asking her to speak to them or to write something. She accepted a job with *Cosmopolitan* magazine and wrote articles about women and flying. Her features encouraged women to try flying.

The next two years were a whirlwind of accomplishments for Amelia, or "AE" as she now preferred to be called. She traveled around the country lecturing on flying. She helped start a new airline offering passenger service between New York City and Washington, D.C. She entered the first Women's Air Derby. She started the "Ninety-Nines" women pilots' organization.

She also bought a Lockheed Vega, which was a faster plane. All the while, she used her earnings to help her family pay off their debts. It was a practice she would continue all her life.

Marriage and a Mission

AE's schedule left little time for socializing, much less dating. She had dated, but never wanted a serious relationship. Part of the reason was that she had seen the troubles in her parents' marriage.

The only person AE did spend a lot of time with was her manager, George Putnam, or GP as he was called. Putnam and his wife had divorced, so GP asked AE to marry him. She said no. He asked again and again. Always she said no. Still, she enjoyed his company and they had many common interests.

"I was interested in aviation, so was he. We both loved the outdoors, books, sports," she said later. "We came to depend on each other, yet it was only friendship between us, or so I thought at first. At least I didn't admit even to myself that I was in love."

Then one day they were at the Lockheed aircraft factory in Burbank, California. AE was waiting for her plane to be warmed up. GP proposed for the sixth time. AE nodded yes, patted his arm and climbed into her plane to fly away. Yet before the wedding, AE asked him to sign an unusual contract. The contract specified that she could retain her career and her independence. It also asked him to let her out of the marriage if they had found no happiness together after a year.

Amelia with her husband, George Putnam.

By 1932, AE was ready for another adventure.

GP agreed to Amelia's contract. The couple was married at GP's mother's home in Connecticut on February 7, 1931. As always, AE wore a simple brown suit and brown shoes to her wedding.

After the wedding, GP quit his work in publishing and took a job with Paramount Pictures. Now he had more time to manage AE's career. He handled her endorsement of a variety of products, from cars to luggage and clothing. He also had arranged for her to endorse a brand of cigarettes after the *Friendship* flight. AE though didn't smoke and thought it was a bad habit. She later stopped endorsing cigarettes.

By 1932, AE felt the time had come for her next adventure. She remembered her promise to herself four years before that she would fly the Atlantic alone. Charles Lindbergh had done it. Now she felt the time had come to show the world that a woman could do it, too.

Lady Lindy

AE quickly began to make plans. She worked with another pilot to prepare herself and her airplane for the flight. The pilot persuaded AE to fly from Newfoundland instead of New York, with an option to land in Ireland instead of Paris. Her airplane received a new engine, extra fuel tanks, and new navigational equipment.

As with the *Friendship* flight, preparations were done in secret. Amelia's husband suggested that she leave on the fifth anniversary of Lindbergh's flight. He always was looking for ways to get maximum publicity. AE conducted her usual rounds of interviews and lectures until 7 p.m. on May 19, 1932. Then she crawled into the cockpit of her plane to begin her biggest adventure yet.

The first four hours of the flight went smoothly. Then she noticed flames peaking through a broken weld in the exhaust pipe of the plane. Next, her altimeter, which tells how high the plane is flying, broke. Then she flew into a storm that lasted an hour.

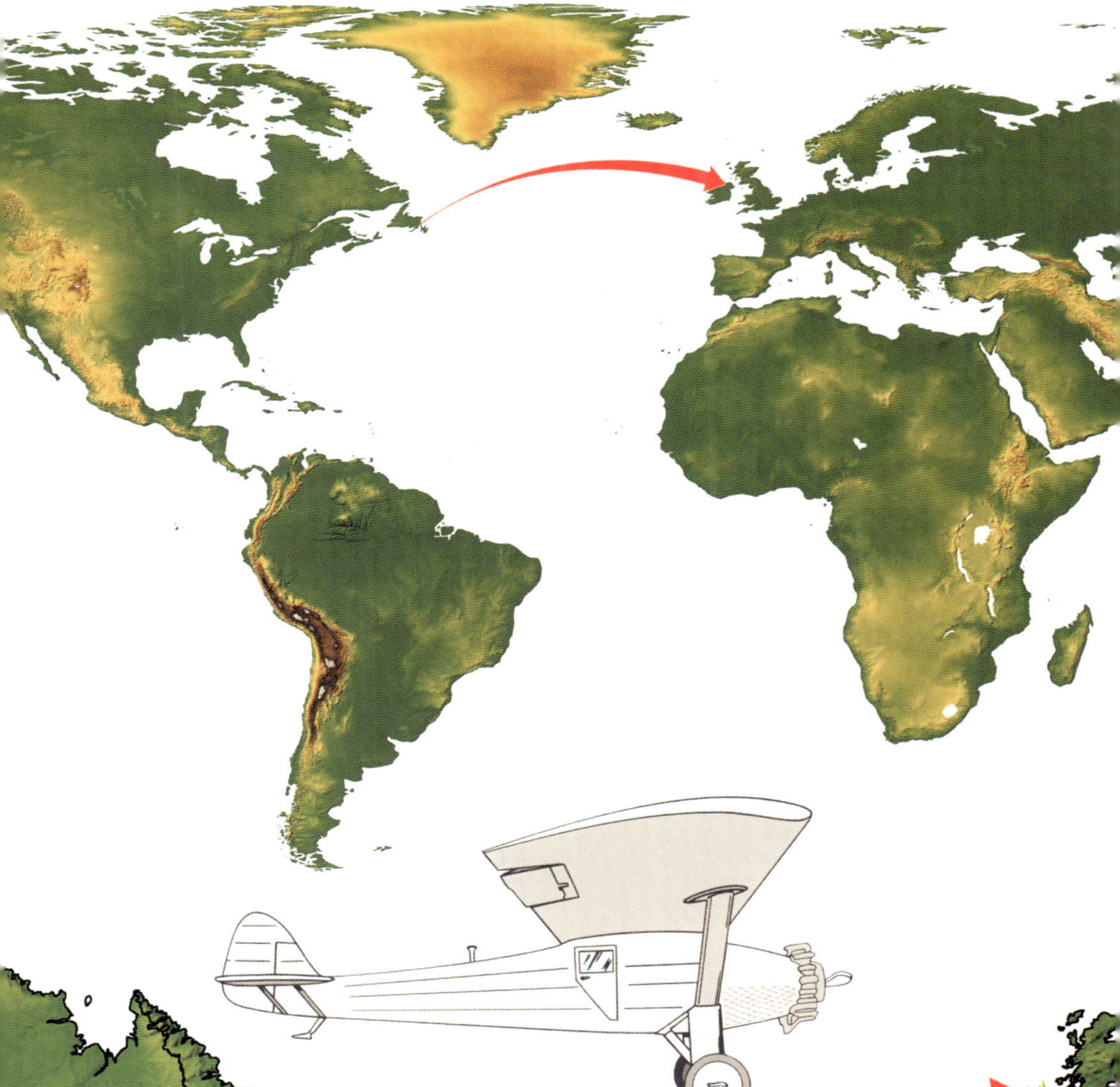

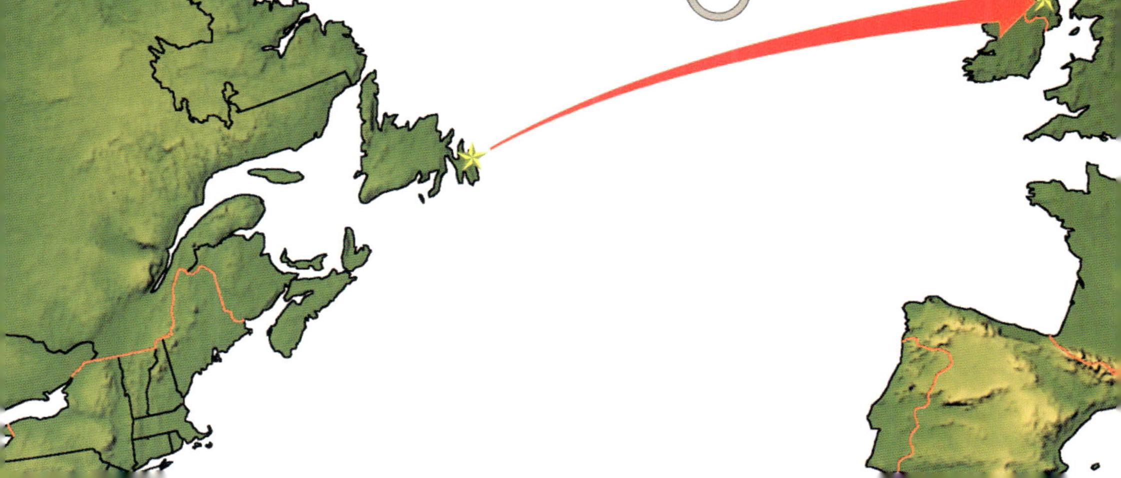

Amelia spent the rest of the flight trying to find the right altitude to fly. Too high and her plane became coated with ice. Too low, where the ice would melt, she ran into fog. With those problems, she decided to seek land as quickly as possible.

AE changed course and landed in a meadow in Londonderry, Ireland. Her flight had taken more than 14 hours and she had flown more than 1,000 miles (1,609 km). On touchdown, she carved out two more permanent records in the history of aviation. She became the first woman to fly a plane across the Atlantic. She also became the first woman to fly across the Atlantic twice.

For the next 12 days in London, AE was treated to a round of celebrations in her honor. She remained modest about her accomplishments. She said she had flown the Atlantic for the fun of it and to show that women were competent pilots. She also wanted to communicate that air travel across the ocean would become available for everyone in the near future.

Amelia is greeted by a happy crowd after landing safely in Ireland.

Crusading for Women

AE's accomplishments won her the Outstanding Woman of the Year award from the *Philadelphia Enquirer* in October 1932. Plus, she continued to win speaking engagements across the country. In one 25-day stretch, AE spoke 23 times for $300 per lecture. More and more, her speeches focused on both aviation and equal rights for women. AE went so far as to deliver a petition for an equal rights amendment to the U.S. Constitution to President Herbert Hoover.

In September 1934, AE teamed up with President Franklin Delano Roosevelt and his wife, Eleanor, to speak at a program in New York City. AE's presentation focused on the lack of opportunities for women in aviation. What few opportunities there were also paid less than those available to men. AE said women and men should be treated equally.

The president of Purdue University in Indiana was in the audience that day. AE impressed him so much he offered her a job. He asked if she would counsel female students at Purdue and advise the school's aeronautics department. AE accepted and planned to begin in fall 1935.

Amelia with First Lady Eleanor Roosevelt.

Later that year, AE and GP's home in New York caught fire. The couple decided to move to North Hollywood, California, instead of rebuilding. They made many friends there, including some movie stars.

Meanwhile, AE had been planning her next flight. So far, no one had flown solo from Hawaii to the U.S. mainland—10 people had tried and died. AE wanted to be the first to succeed. She loaded her plane on a ship bound for Hawaii. On December 22, 1934, she left Honolulu for California.

The flight went smoothly. Eighteen hours after leaving Hawaii, she landed to the cheers of a crowd in Oakland, California. Two more world records were hers. She was the first person to fly solo across the Pacific. She also was the first person to fly solo across both the Pacific and Atlantic.

AE was now more popular than ever. The president of Mexico invited her to fly there. The Mexican government even printed a postage stamp about her visit. GP had AE sign some of the stamps and then sold them to collectors to help finance the flight. She made the flight from Burbank, California, to Mexico City in a record 13 hours and 32 minutes.

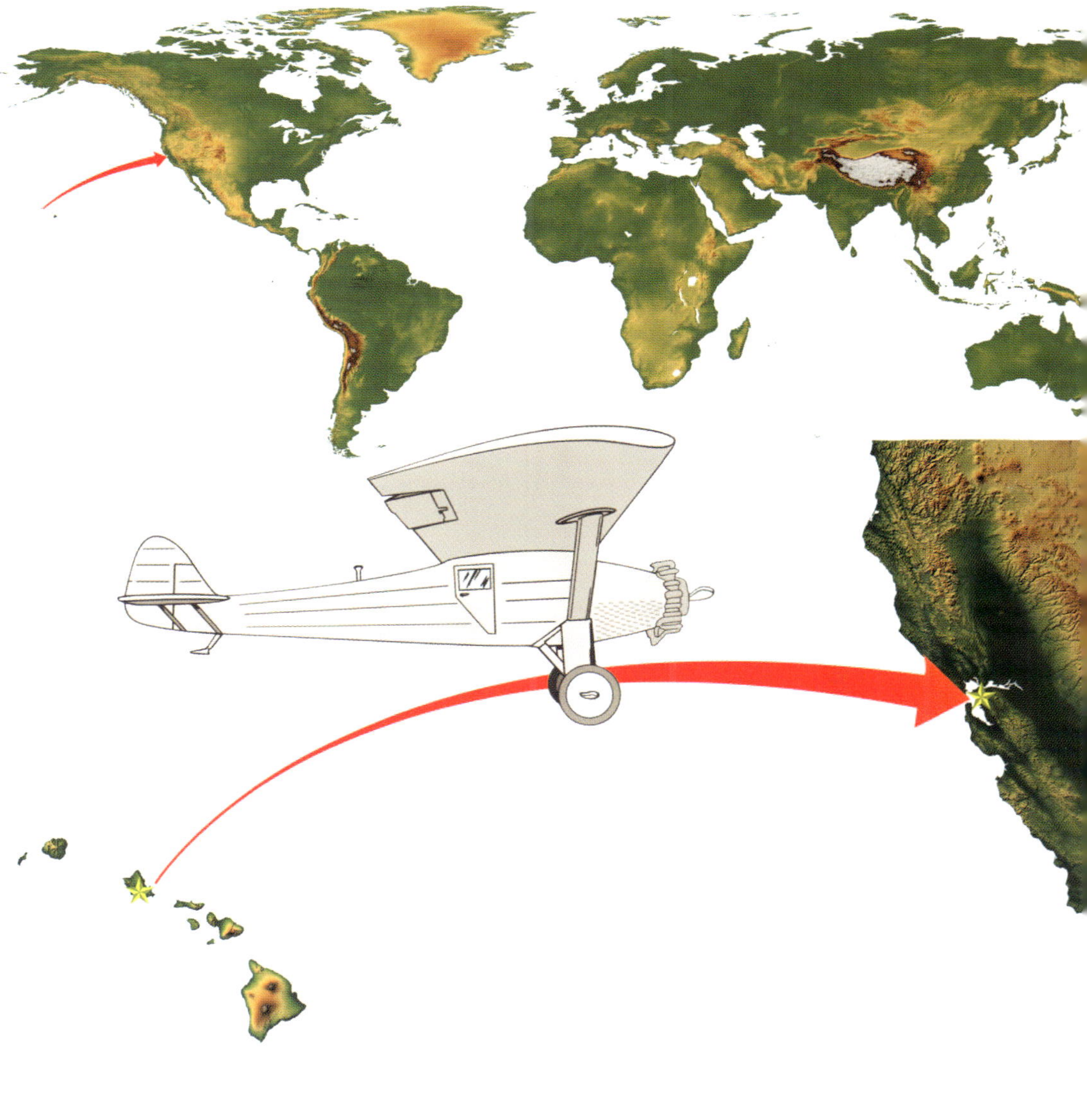

Last Grand Adventure

The next two years were a whirlwind of lectures for AE. She also took up her counseling job at Purdue University. While at Purdue, GP encouraged the school to fund a flying laboratory for his wife. With private donations, AE bought a two-engine Lockheed Electra monoplane. The university thought AE could use it for research for the school's aeronautics department. AE had something else in mind.

She had been considering another long-distance flight for some time. Now she decided she wanted to tackle a flight around the world at the equator. Other pilots had flown around the world, but none had taken this, the longest route.

As before, AE and GP went into action. GP made the necessary arrangements for the flight, including asking the U.S. to build a runway on Howland Island in the middle of the Pacific Ocean. He got landing permits from the various countries where she would stop. He worked to finance the venture by setting up more lectures for AE.

Amelia Earhart preparing to fly around the world.

AE worked with other professionals to turn her airplane into a state-of-the-art flying machine. She practiced flying by using instruments only. Sadly, she was unable to spend much time practicing in her new airplane.

Amelia Earhart and Fred Noonan load their plane for a world-spanning flight.

AE announced her proposed flight to the world on February 12, 1937. She had been so busy preparing for the event that she had forgotten her own wedding anniversary. Originally, AE planned to fly westward, beginning with a flight to Hawaii. She would be flying with a navigator, Harry Manning, and a back-up navigator, Fred Noonan. They took off for Honolulu on March 17, 1937. They arrived without incident. Bad weather then delayed their take off for Howland Island for two days.

On March 20, they tried to fly the next leg to Howland Island, but the plane had trouble on take off and was badly damaged. It had to be taken back to California for repairs. AE remained undaunted. She vowed to try again. GP went to work raising the more than $50,000 it would take for another attempt.

Meanwhile, plans for the flight were changing. The navigator, Harry Manning, was a Navy pilot whose leave was nearly over. He had to return to his ship, leaving Fred Noonan as the navigator and backup pilot. Noonan had worked for Pan American Airlines before a drinking problem cost him his job. He promised AE he would not drink during their flight. Also, weather problems spurred AE to change her flight plans. Now she would fly eastward, beginning at Miami, Florida.

Around the World

On June 1, 1937, AE and Noonan took off from Miami bound for Puerto Rico. They planned to fly by day and land each night for refueling and maintenance. Their route included stopovers in Venezuela, Dutch Guiana (now called Surinam), Brazil, Africa, Pakistan, India, Singapore, Indonesia, Australia, and New Guinea. The last legs of the trip would be to Howland Island, and then on to Hawaii and California.

GP had done an excellent job organizing their layovers. Wherever they landed, fuel was waiting in barrels labeled with AE's name. Airplane parts also were available. AE and Noonan were greeted warmly wherever they went and treated to banquets in their honor. AE wrote in her journal that she didn't like the hectic pace of the flight. "We're always pushing through, hurrying on our long way... instead of enjoying the place... . A situation, alas, about which there is no use complaining. After all, this was not a voyage of sight-seeing."

After nearly a month of traveling, they arrived at Lae, New Guinea. It was June 29. They had traveled more than 20,000 miles (32,186 km). They had some 7,000 miles (11,265 km) yet to go, and all of it was over water. AE hoped to be home by July 4 to speak at a Fourth of July engagement GP had arranged.

Weather delayed their flight to Howland Island. When they could leave, both AE and Noonan looked gaunt and tired. The trip had taken a heavy toll on them both. Some people were afraid Noonan had begun drinking again. AE seemed to be running on sheer determination. She and Noonan took off the morning of Friday, July 2.

The flight to Howland Island would be the most grueling leg of all. AE had estimated the flight time around 18 hours. All of it was over the ocean. Howland Island was a mere mile and a half long. Any error in navigation could be disastrous. The Coast Guard cutter *Itasca* was waiting at Howland to to help guide AE via radio or, if necessary, Morse code. They did not know AE had left the special radio and Morse code equipment in Miami, thinking she would not need them.

At 2:45 a.m. on July 2, the *Itasca* began to pick up AE's radio transmissions. Yet, since she had not brought all her radio equipment, she could not hear their responses. She continued to transmit on and off through the night but heard no response. Her last transmission came at 8:44 a.m. Breathlessly and shrilly, she said, "We are on line of position 156-137. We are running north and south."

The *Itasca* crew thought the message meant that AE knew she should be near Howland Island, based on how far east she had flown. However, she could not find the island. She was flying north and south at that longitude, trying to find it.

No one ever heard from Amelia Earhart again.

Amelia Earhart helped bring opportunities to female flyers.

Missing

*J*ust over one hour after AE's last transmission, the *Itasca* crew decided AE must have ditched her airplane at sea. The ship estimated she was some 35 to 100 miles (56 to 161 km) off the coast of Howland Island. If she had been unable to make land, the plane's empty fuel tanks could have kept the plane afloat for a short time.

President Roosevelt authorized a full naval search. For two weeks, nine ships and 66 aircraft combed the region of her last known position. They covered more than 250,000 square miles (647,500 square km) of the Pacific. The effort cost more than $4 million.

The Electra had carried a life raft, yet no raft ever was found. Nor did searchers discover any airplane wreckage. Some people theorized the plane had gone down in territory controlled by the Japanese. Japan had forbidden the U.S. Navy to search there.

Amelia Earhart

GP continued the search even after the Navy gave up. Yet by October, even he had abandoned any hope of finding his wife alive.

Obviously something had gone radically wrong. The static on the radio indicated the plane had flown into stormy weather. A Navy plane flying near Howland Island had run into bad weather the day the Electra disappeared. With cloudy conditions blocking the stars, Noonan could easily have gotten off course. There might have been trouble with their navigation instruments. It was clear they had problems with their radio equipment.

Amelia at the controls.

Amelia during her first parachute jump in 1935.

Unsolved Mystery

Amelia Earhart's fate remains a mystery. Many people claimed to have seen her. There are many theories about what happened to her as well.

Some people say AE was on a spy mission for President Franklin D. Roosevelt when she tried to fly around the world. Her job was to spy on Japanese activity in the Pacific area. Others believe the Japanese captured Earhart when her plane went down. Some say she was the voice of "Tokyo Rose," who broadcast to American soldiers during World War II. Still others think she lived for years on a South Pacific island with a native fisherman.

In 1988, The International Group for Historic Aircraft Recovery (TIGHAR) launched a special investigation. They wanted to determine once and for all what had happened. The search would be the most exhaustive since that mounted by the Navy in 1937.

Amelia is questioned at an airfield press conference.

Amelia at a tickertape parade held in her honor.

TIGHAR researchers made four trips to the South Pacific island of Nikumaroro. They found the remains of a woman's shoe dating from the mid-1930s, which matched the size and style worn by AE. They also found evidence of a campfire.

Their findings match other bits of evidence. On July 9, 1937, a Navy searcher noted signs of recent habitation on the island. The searcher did not know the island was supposed to be uninhabited at the time. In 1940, bones also were found on the island, along with a bottle and a nautical sextant. Aircraft wreckage showed up on the island more than 10 years later.

TIGHAR researchers theorize that AE landed her airplane on the reef flat of the island and swam ashore. The aircraft then sank. In 1971, an aircraft engine similar to the kind found on AE's plane was found in the area. Researchers think it was washed up on shore by a storm. TIGHAR continues to search for the rest of the plane in the waters surrounding the island.

AE's Legacy

*D*espite all the theories and searches, the world may never know for certain what happened to AE and Fred Noonan. What the world does know is the impact AE's life had on the world.

Had she lived, who knows what AE might have gone on to do. She likely would have continued to crusade for women's rights. She probably would have looked for more challenges in the air and on the ground. She had hoped to return to her job at Purdue after the around-the-world flight and fly only for pleasure from then on.

In one of her last letters to her husband, she wrote about her decision to undertake her around-the-world trip. "Please know I am quite aware of the hazards," she wrote. "I want to do it because I want to do it. Women must try to do things as men have tried. When they fail, their failure must be but a challenge to others."

Certainly, AE inspired other women to undertake their own challenges. Her memory continues to inspire other women of the world today.

Amelia Earhart

Amelia Earhart inspired women to challenge themselves.

Timeline

July 24, 1897: Born in Atchison, Kansas. Daughter of Edwin and Amy Earhart.

1916: Graduated from Hyde Park High School in Chicago.

1917: Became nurse's aid at Spadina Military Hospital, Toronto.

1922: Earned pilot's license after taking flying lessons inspired by a 10-minute plane ride.

1928: First woman to fly across the Atlantic Ocean.

1931: Marries George Putnam.

1932: First woman to fly solo across the Atlantic Ocean.

1934: First person to fly solo across the Pacific Ocean. Also set record for first person to fly solo across *both* the Atlantic and Pacific Oceans.

1935: Starts work at Purdue University, advising the school's aeronautics department and counseling female students.

1937: While attempting to circle the globe in an airplane, Amelia is lost somewhere over the Pacific Ocean, never to be heard from again.

Where on the Web?

http://www.ellensplace.net/ae_intro.html

Amelia Earhart biography.

http://www.ameliaearhartmuseum.org/

Amilia Earhart birthplace museum.

http://www.ameliaearhart.com/

The official site of Amilia Earhart.

http://www.womeninaviation.com/home.html

Women in Aviation.

http://www.history.navy.mil/faqs/faq3-1.htm

Naval Historical Center, Dept. of the U.S. Navy.

http://foia.fbi.gov/earhart.htm

F.B.I. documents obtained through the Freedom of Information act relating to Amelia's disappearance.

Glossary

altitude

The distance of a plane above the ground when flying.

biplane

An airplane with two wings placed one above the other.

bloomers

Short loose pants gathered at the knees.

dormitory

A building with sleep rooms, often on college campuses.

instruments

Devises used to navigate airplanes.

Morse code

A special way to send messages across short distances.

pontoons

Floats on an airplane.

sextant

A navigational instrument.

spin

When a plane dives toward the ground while spinning around.

stall

When a plane flies with the nose up. The wings stop producing lift, causing the plane to plunge toward the ground.

Index